THE

LITTLE BOOK OF

Chickens

THE
LITTLE BOOK OF

Chickens

An illustrated guide to the
extraordinary lives of chickens

JESSICA FORD

ILLUSTRATED BY AMY HOLLIDAY

WILLIAM
COLLINS

William Collins
An imprint of HarperCollins*Publishers*
1 London Bridge Street
London SE1 9GF
WilliamCollinsBooks.com

HarperCollins*Publishers*
Macken House, 39/40 Mayor Street Upper,
Dublin 1, D01 C9W8, Ireland

First published by William Collins in 2025

A catalogue record for this book is available from the British Library
Library of Congress Cataloging-in-Publication Data has been applied for.

ISBN 978-0-00-875343-6

Text: Jessica Ford
Cover and interior illustrations: Amy Holliday
Design: Jacqui Caulton and e-Digital Design
Copyedit: Helena Caldon
Proofread: Rachel Malig
Project editor: Caitlin Doyle
Production controller: Jayoti Shah

Printed and bound in Bosnia and Herzegovina

MIX
Paper | Supporting
responsible forestry
FSC™ C007454

This book is produced from independently certified FSC™ paper
to ensure responsible forest management.

For more information visit: www.harpercollins.co.uk/green

Dedicated to Patti Bearpaw,
whose love of birds inspired generations.

CONTENTS

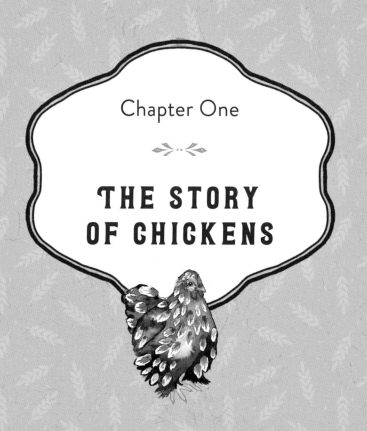

Chapter One

THE STORY
OF CHICKENS

THE HISTORY OF CHICKENS

There are very few animals that match the humble chicken for their importance to humanity. Prolific, exotic, easy to handle, and quite capable of fending for themselves, chickens have been an integral part of human survival and culture for thousands of years.

CHICKEN ORIGINS

Chickens are believed to have been domesticated 8,000–10,000 years ago in modern-day Southeast Asia. They are primarily descended from the Red Junglefowl (*Gallus gallus*), and possibly the Grey Junglefowl (*Gallus sonneratii*) and the Green Junglefowl (*Gallus varius*) as well, all of which are wild species that are still in existence today. The wild populations are generally stable, and flocks are easy to find in their native range, though some are threatened by hybridization with domestic chickens.

The Junglefowl and the modern chicken share many traits, but thousands of years of domestication has led the modern-day chicken to be fundamentally distinct from its wild counterparts. The differences can vary by breed, but in general, modern chickens are larger, slower, and far more friendly to humans than Junglefowl. Chickens are also better egg-layers, but generally poorer fliers compared to the wild Junglefowl, a species capable of some sustained flight.

Although best known as a source of meat and eggs, chickens have played many roles throughout human history. In fact, the earliest depictions of domesticated chickens do not revolve around food production at all. According to fossil records dating back to about 5400 BCE, the earliest chickens were primarily used in religious and burial offerings. Starting at about 2500 BCE, domesticated chickens expanded west, brought by travelers through present-day Pakistan and Mesopotamia. They later appeared in ancient Greece and Egypt before eventually making their way to Europe via the Silk Road. Chickens spread further east as well, joining Polynesian voyagers across the Pacific.

These early chickens were not the heavy-set, common backyard breeds that we know today. They were far smaller and flightier, and not particularly good sources of food. Their eggs were likely considered a delicacy, and they were probably more valued for their exotic appeal, religious symbolism, and—starting in about 2000 BCE—for their entertainment value in cockfighting.

It wasn't until about 500 BCE, when ancient farming techniques saw a jump in innovation and efficiency, that chickens became more of a widespread food source. At this point archaeological records show fascinating advances in chicken breeding, incubation, and husbandry to develop breeds that were heavier for meat and egg production. Once established as an easy livestock for food, the humble chicken's popularity surged to meet the protein demands of growing human populations.

HEN FEVER

The chicken's development in human history saw another surge during the Victorian era of the mid-nineteenth century, with the onset of "Hen Fever." This trend, which coincided closely with the similar "Dog Fancy" trend, started in 1840 when Queen Victoria was gifted a flock of exotic Cochins (see image on previous page) from the Emperor of China. Her love of the birds spurred a chicken-breeding craze across Europe and the U.S. that led to the development of dozens of new chicken breeds, and launched competitive poultry shows and breeders' clubs. The Hen Fever craze became so popular that it experienced its own economic bubble, which eventually burst after about 30 years when the supply of fancy chickens outweighed demand. The impacts of this trend have lasted for over 150 years, with many Hen Fever-era breeds and poultry shows continuing throughout the world.

Today, the chicken reigns supreme as the primary animal food source throughout the world, with over 26 billion meat- and egg-production chickens alive at any given time. They are also rapidly regaining favor as backyard companions, recently earning their place as the third-most-popular "pet" in the United States and the United Kingdom.

FUN CHICKEN HISTORY FACTS

1 In ancient Greece, chickens were considered sacred to gods like Hermes and Ares.

2 The "chicken or the egg" question dates back to ancient Greek philosophers. Aristotle himself once debated which came first.

3 In ancient Rome, chickens were believed to be omens that could predict the results of a battle.

4 Chickens are direct descendants of theropod dinosaurs, and they are the closest living relative to the T-Rex!

5 Chickens play a symbolic role in many ancient religions, including Buddhism, Judaism, and Zoroastrianism.

6 Many cultures believed that chickens could predict the seasons and the weather. In fact, many today still believe that chicken behavior is a good indicator of the change of seasons. When older hens restart egg laying after a winter break, for example, it is considered a sign of when spring will arrive.

7 The phrase "don't count your chickens before they hatch" dates back to *Aesop's Fables*, written around 600 BCE.

CHICKEN ANATOMY

A chicken's anatomy is surprisingly different from that of mammals such as dogs and cats. Although not the best fliers, they are true birds and share birds' unique traits and anatomy.

For starters, a chicken's respiratory system is significantly more complicated and efficient than a mammal's. They have lungs, but they also have air sacs throughout their bones, which are hollow. This special ability to "breathe" straight into their bones not only makes them lighter, it also allows them to generate enough oxygen throughout their bodies for flight, which is a very energy-intensive activity. This enhanced respiration comes with a price, though, because like all birds, chickens are extremely sensitive to air quality, and should be protected from aerosol sprays, cedar wood, smoke, and mold, especially in closed spaces.

Like many other birds, although chickens have nostrils—called nares, which are located at the very top of the beak—they do not have a particularly good sense of smell. Instead, their primary sense is their eyesight, which is exceptional. Chickens can see much better than humans can, including colors that we cannot perceive, like ultraviolet light, and patterns of movement too fast for our eyes to notice. Each of their eyes also has the unique ability to focus separately at the same time: one eye is always far-sighted to better perceive predators, while the other is near-sighted to better find food. Chickens' eyes are also proportionally huge, taking up about 50 percent of their skull, which gives them over 300 degrees of vision. Because vision is so important to their survival, chicken eyes have extra protection with a third eyelid—a translucent membrane that helps them see while keeping their eyes safe from dirt or other debris. Despite their amazing eyesight, chickens do not have good night vision, and are therefore very vulnerable to predators after dark.

Another unique bird trait that chickens have is an avian digestive system that is believed to be similar to that of some dinosaurs. Since birds do not have teeth, they cannot chew.

Instead, chickens and other birds swallow morsels of food whole and store them in a special pouch called a "crop." This crop is very useful, allowing them to fill up on food and then digest it slowly, especially overnight. Once through the crop, the food is pushed into the "gizzard," which is similar to a mammal's stomach but it also holds small rocks that chickens ingest periodically to help break down food. From there, the food is absorbed through the intestines and eventually excreted as waste through the cloaca, along with urine, which is excreted at the same time. It is a highly efficient digestive system that supports chickens' fast metabolisms.

CHICKEN FEATHERS

Chickens might not have the ability to handle sustained flight (though some breeds can fly 50 feet/15.5 meters or more), but feathers are still a very important part of what makes a chicken a chicken. A chicken's feathers are not just nice to look at; they play a vital role in their health, in protecting them from the elements, in attracting a mate, and especially in maintaining body temperature.

Feathers are made of keratin, like human fingernails, and are mostly hollow to be as lightweight and insulating as possible. Each feather consists of a main "sheath" and hundreds of small "filaments", which connect together with tiny hooks. There are many types of feathers found in chickens, including down feathers, which are small and fluffy like an undercoat, and rigid flight feathers on the wings to provide lift for short bursts of flight. All feathers must be meticulously groomed and maintained, which all birds do in a process called "preening." A healthy chicken will preen their feathers throughout the day, using a special gland above their tail to coat themselves in oil with their beaks. This keeps their feathers waterproof and weather tight.

Chicken feathers come in a dizzying range of colors and patterns thanks to centuries of selective breeding, making them among the most beautiful and diverse of bird species. Roosters especially boast special hackle, saddle, and sickle feathers, which are used to show off and attract mates.

Chickens carry mites that can be passed on to humans.
False! While feather mites and lice can occur occasionally in chicken flocks, they are not compatible with humans or other mammals and die quickly without a bird host. Chickens kept in a healthy, clean environment will not experience mites as a major, ongoing issue.

Chickens are smelly.
False! Chickens are not inherently smelly animals. The birds themselves have a mild dusty smell from their feathers and frequent dust baths. Their coop should also have minimal odor as long as it is properly cleaned and maintained.

Chickens attract mice.
False! Mice are attracted to spilled feed, not chickens. Proper food storage and low-mess feeders will prevent mice and rats from becoming an issue.

Chickens lay year-round.
False! Hens naturally lay in seasonal cycles. They may slow down or cease laying while molting, brooding, or during the winter months.

Chickens can't be tamed or trained.
False! Chickens can recognize the faces of their owners, and some will actively seek out affection from their favorite humans. Chickens are also highly food motivated and are easily trained to perform tricks.

Chickens are only allowed in rural areas.
False! Many cities, towns, and suburban neighborhoods now allow small flocks of chickens, and the trend is expanding every year. Keep in mind that most neighborhoods will have strict rules, such as the number of birds, coop location, and permits.

CHICKEN ANATOMY GLOSSARY

EAR LOBES
A soft patch of skin that helps protect the ears. They can come in red, blue, or white.

SADDLE FEATHERS
The long, narrow feathers above the tail in roosters.

SICKLE FEATHERS
The long, curved tail feathers in roosters.

PREEN GLAND
A small lump above the tail, which excretes a special oil that birds use to keep their feathers in good condition.

CLOACA
The single opening below the tail, also known as the "vent." All mating, egg laying, and excrement happens through this single opening.

SHANK
A chicken's leg. Usually scaled, but sometimes also feathered in some breeds.

COMB

The fleshy crest on the top of a chicken's head. The comb can be "single" (thin), but it can also come in "walnut," "rose," and "pea" shapes, among others. The single (thin) comb is the most common. Thought to be used as a temperature regulator—especially in chicken breeds that originate from hot climates—as well as a vitality signal to other chickens.

NARES

A chicken's "nose," which sits on the top of the beak.

WATTLES

The fleshy parts beneath a chicken's beak, which are similar in purpose and color to the comb.

CROP

The pouch where food is held right after eating. Appears as a large lump on one side of the chest when full.

HACKLE FEATHERS

The neck feathers, particularly on roosters. These, in some types of chickens, can be long, fine, and brightly colored. Roosters will stand these up to look more intimidating.

KEEL

The breastbone of a bird, which feels like a ridge of bone down the chest toward the belly.

FLIGHT FEATHERS

The large wing feathers that chickens use for short bursts of flight.

TOES/CLAW

Most chickens have three claws pointing forward and one backward, called the claw or spur. These are used for scratching and balance.

CHICKEN ANATOMY FACTS

1 A rooster's hackle (neck) feathers are often used for making fishing lures.

2 The color of a hen's ear lobe is said to determine the color of egg she will lay (this is only true in some breeds, such as the Leghorn and Rhode Island Red).

3 You can always tell how "full" a chicken is by feeling how large their crop is.

4 Male chickens do not have external genitalia, making them difficult to sex as chicks.

12 BEST BACKYARD CHICKEN BREEDS

One of the most fascinating and wonderful aspects of the backyard chicken is its diversity. While the precise number of breeds is open to debate, there are certainly hundreds known in the world today, in all kinds of shapes, sizes, colors, patterns, and egg-laying ability. With so many to choose from, selecting a breed can be a daunting task, especially for a beginner or those with limited space. If you are looking for a shortlist of tried-and-true chicken breeds, the 12 breeds on the pages that follow have been selected for being especially well-suited to backyard flocks for a range of reasons that I have elaborated on.

ORPINGTON

The fluffy, docile Orpington is one of the most popular backyard chicken breeds, and for good reason! Fondly referred to as the "Golden Retriever of the chicken world," this breed is famous for its gentle temperament and affectionate nature. If you have young children, are new to chickens, or want a reliable layer and quiet breed, the Orpington is a great choice. They even tolerate handling well and tend to be gentle with people and other pets.

Developed in England during the 1800s, the Orpington was bred to be a hardy and functional homestead breed, able to withstand cold weather and provide families with both meat and eggs. Orpingtons were imported into the U.S. early on, where the breed became a fast favorite, and it has remained so ever since. Although the Buff Orpington (see right) is by far the most popular variety, this breed comes in many other colors, too, including Black, Lavender, Blue, Jubilee, and Chocolate. Today, this breed is available in English and American variations, with the English being a little more stocky and fluffy. All colors and variations are a good choice for backyard flocks.

While known for being especially gentle, this breed is no pushover in its flock. Thanks to their large size, the Orpington is not often bullied, and the roosters are good protectors to their hens. Hens are decent layers of medium-sized pink–brown eggs, but are known to go broody frequently. They make fabulous mothers, often adopting others' eggs and chicks. Orpingtons are reliable foragers out on a sprawling homestead, but they are perfectly content in smaller enclosures too. Being a heavier breed, they are not as likely to jump over fences.

PLYMOUTH ROCK

The Plymouth Rock is one of the United States' oldest heritage breeds, and it has been a dependable member of homestead flocks for generations. This breed was by far the most widely kept in the United States for decades throughout the twentieth century, and for good reason! The Plymouth Rock is a beginner-friendly, hardy bird that is an exceptional layer.

This breed was developed in early nineteenth-century Massachusetts by crossing the older Dominique with Black Javas and other breeds. They've been known by the nickname "America's Favorite Breed" ever since. The Plymouth Rock's most popular color by far is the black-and-white Barred variety (see left), and the breed is commonly called the "Barred Rock" for that reason. However, this breed comes in other colors too, notably white, blue, and black.

Plymouth Rocks are known for their intelligence and outgoing nature. Like the Orpington, they are rarely bullied and are generally friendly toward other members of their flock. Roosters are fiercely protective of their flock, though, and are known to be aggressive. The hens are calm, quiet, and very adaptable. They are excellent layers of large brown eggs. While they do go broody and make great mothers, Plymouth Rock hens are not overly prone to broodiness. They thrive in large free-range flocks, small backyard spaces, and everything in between, making them one of the most adaptable breeds you can own.

RED AND BLACK SEX-LINKS

Sex-links are not a true breed, but are a production hybrid developed for egg laying. They get their name from their unique attribute, where female chicks are easily distinguishable from males by their color differences. Each type of sex-link will have its own unique name, based on the hatchery that developed it. Common Red Sex-links include the Cinnamon Queen, ISA Brown, Red Star, Gold Star, and Golden Comet.

The breeds used to develop Red Sex-links (see right) can vary, but they are generally a mix of Rhode Island Red or New Hampshire Red with Plymouth Rock, Delaware, or sometimes White Leghorn. Because these "breeds" are technically hybrids, they do not breed true, meaning the offspring of two sex-links will not be another sex-link.

Sex-links are not only prized on industrial egg farms, they also make beloved backyard pets and homestead birds. Their feed-to-egg ratio is impressive, and each hen can lay 300 large brown eggs or more per year. They are also known for being quite friendly, especially toward humans. It's worth noting, however, that these hybrid breeds tend to develop health and reproductive issues frequently, and are known to have a shorter lifespan than other hardier breeds, averaging about two to three years.

In addition to Red Sex-links, Black Sex-links, also known as Black Stars, are available, and have similar characteristics.

RHODE ISLAND RED

If you own a homestead or small farm, you can't go wrong with the Rhode Island Red. This breed is easily a top three choice, and has been for generations. Like the Plymouth Rock, the Rhode Island Red was developed in New England during the early nineteenth century to be a hardy, dual-purpose farm bird. They became so popular and treasured for their dependability, they actually became the state bird of Rhode Island—the only domesticated bird to earn the title.

Thanks to their many excellent characteristics, the Rhode Island Red is a foundation breed for many other breeds and hybrids, including those bred for meat and eggs at an industrial scale. They are a beautiful breed, with glossy dark mahogany feathers and yellow legs. Although less common, you can also find this breed in pure white—usually sold as the Rhode Island White.

This adaptable breed does well in smaller spaces, but they truly thrive when allowed to free range. They are outgoing, intelligent, and excellent foragers. Rhode Island Reds are also exceptionally friendly toward their keepers, and can be quite affectionate. They are known to be assertive and sometimes aggressive toward other poultry, however, especially if they are not given enough space to roam and roost. This is one of the more hardy breeds you can find, and they are known to live long lives of 8 to 12 years.

AMERAUCANA

The Ameraucana is a delightful heritage breed of chicken that is prized for its fluffy face, sweet nature, and especially its bright blue eggs. Although commonly confused with Easter Eggers and similar-colored egg crosses, the Ameraucana is a true breed, and one that comes with dependable looks and egg color.

The Ameraucana (not "Americana," which is an Easter Egger) is a descendant of the enigmatic Araucana chicken—a unique breed from South America that is both rare and prone to hatching issues. American fanciers in the 1970s sought to develop a new, hardier Araucana chicken breed while maintaining its fluffy cheek muffs and pretty blue eggs. The resulting Ameraucana quickly grew in popularity. Note: the Ameraucana is still considered part of the Araucana breed outside the U.S.

Ameraucanas come in many different color varieties, including black, blue, lavender, and partridge. All varieties exhibit fluffy beards and muffs, small pea combs, and of course those desirable blue eggs. This breed is very hardy, especially to the cold, and is known for its gentle, sweet temperament around both humans and other chickens, making it an excellent choice for beginners. Ameraucanas are decent layers, averaging about 200 eggs per year.

BANTAM COCHIN

The delightful Pekin—known as the "Bantam Cochin" in the U.S. and Canada—is an adorable chicken breed that is quickly gaining popularity in modern backyard flocks. These fluffy little birds sport feathered feet, round rumps, and sweet personalities.

The Cochin is an ancient breed originating from the Cochin region in China, with records of them dating back over 1,000 years. They arrived in Europe in the nineteenth century. Their unique looks quickly gained the favor of Queen Victoria, and helped shape the rise of Hen Fever during the Victorian era. In North America, the Pekin is considered a bantam size of the Standard Cochin. In the U.K., however, they are considered a standalone breed with no other size counterpart.

Regardless of what they are called, Bantam Cochins are sweet-natured, quiet, and simply lovely to be around. They are especially well suited to small backyards, thanks to their smaller space requirements and tendency to be quieter than other breeds. They are not particularly good layers, but they do make excellent mothers. If you acquire any Bantam Cochins, keep in mind that their feathered feet need to be kept dry and clean, and they may become targets for bullying if kept in mixed flocks with larger, more aggressive breeds.

D'UCCLE BANTAM

Once an uncommon breed outside the chicken fancy world, the D'Uccle Bantam is now receiving some much-deserved attention as a backyard pet. This diminutive breed features fluffy beards and muffs, feathered feet, and some remarkable feather patterns. They are also curious and friendly, and a guaranteed conversation starter if you own them.

The Barbu d'Uccle, or Belgian d'Uccle, was developed in the late nineteenth century outside Brussels during the height of Hen Fever. The breed was introduced to the rest of Europe and the U.S. shortly after, where it quickly became a recognized breed. The D'Uccle is a true ornamental bantam breed, with no standard-sized breed counterpart, and no functional purpose on a farm. They are simply meant to be beautiful pets or show birds. Because of this intended purpose when the breed was developed, D'Uccles are quite personable and tend to enjoy being handled.

Despite their size, D'Uccles are decent layers of small, white eggs. They initially only come in the spotted mille fleur color pattern (see right), but are now available in many colors, including porcelain, white, blue, and splash. If you have a smaller backyard, have small children, or simply want to add a unique breed to your flock, the D'Uccle is a fabulous choice.

SAPPHIRE GEM

The stunningly pretty Sapphire Gem is a fantastic chicken breed if you are looking for an excellent layer with unique plumage. This breed, like other sex-links, is technically a hybrid. They have become very popular in recent years, thanks to their large brown eggs, silver feathers, and gentle demeanor, and deserve their own shoutout as a backyard chicken breed.

The Sapphire Gem is a more recent hybrid breed with a somewhat uncertain history. They are believed to have originated in the Czech Republic, developed from the Blue Plymouth Rock. This breed is now a trademarked hybrid from Hoover Hatchery in the U.S., which is the sole source of the Sapphire Gem. Sapphire Gems closely resemble the rarer Blue Andalusian, but it is unclear if the two breeds are at all related. A Sapphire Gem hen will lay up to 300 eggs per year, placing this breed firmly in the top-10 best laying breeds.

Sapphire Gems tend to be more heat-hardy than most breeds, and are therefore an excellent choice for warm climates. They also tend to be a little more docile around other chickens than other industrial layers, and make great beginner-friendly additions to the flock. This is another great breed to have around children, or in small backyards, but they are intelligent and savvy enough to free-range on sprawling homesteads as well.

COLORED EGGER

(Easter Egger, Olive Egger, etc.)

Colored "Egger" chickens go by many different names, depending on the cross and desired egg color. They are technically not recognized breeds, but many have been in development for decades. The goal of these breeds and mixes is to achieve beautiful, unique egg colors, or specific egg colors mixed with another trait. While their appearance can range widely, all colored eggers tend to be mild-mannered and intelligent, and make wonderful backyard pets.

Some colored eggers that you may find at hatcheries or breeders include Easter Eggers (sometimes sold as "Americanas," see right), Olive Eggers, Moss Eggers, Fibro Easter Eggers, Silked Easter Eggers, and Azure Eggers. The breeds used to develop these may vary, even within the same egger "breed." They often involve one breed with the blue egg gene, such as the Ameraucana or Cream Legbar, and one breed with a brown egg gene, such as the Marans or Welsummer. The resulting mixes tend to lay delightfully unique eggs in a multitude of colors, ranging from bright blue to olive green and everything in between.

Because colored egger "breeds" are incredibly diverse and unstandardized, their egg color is not guaranteed, so it's best to get at least several at a time or stick with a true breed if you are looking for a specific egg color. Colored eggers are generally very good layers of 250+ eggs per year, and overall are healthy and long-lived. Depending on their breed parentage, though, they can be flighty and sometimes a little noisy, but they are known to be a good choice for beginners. If you have children, a colored egger of any kind is an excellent choice.

AUSTRALORP

If Orpingtons are the Golden Retrievers of the chicken world, Australorps are certainly the Black Labradors. They are overwhelmingly popular among homesteaders and backyard keepers alike for their calm demeanor with both people and other chickens, dependable egg laying, and hardy beauty. If you are new to chickens and don't know what exactly you want out of a flock, you can't go wrong with an Australorp—they make excellent backyard pets and are good for beginner keepers of all ages.

The Australorp was developed from early Black Orpington lines in Australia (hence their name). Breeders wanted a hardy chicken like the Orpington, but one that was better suited to Australia's hotter, drier climate. The resulting Australorp is hardy in heat as well as cold, takes confinement well, such as in chicken runs or fenced pens, and shares the Orpington's docile temperament. Black Australorps are almost always found in pure, iridescent black (see left), but some rare varieties do come in white and blue.

Australorps are generally excellent layers of large brown eggs, but they are known to go broody often. Hens make excellent mothers. Roosters are known for being fierce protectors, and their black feathers are rumored to help deter hawks, which confuse them for ravens or crows.

SILKIE

If you are looking to get chickens for companionship, a Silkie is arguably the best choice. This unique, ancient breed is famous for its gentle, affectionate nature, and extra-soft, cuddly, fur-like feathers. They are not good layers, but they are among the most gentle-natured breeds of poultry.

The Silkie is a fascinating breed with unique attributes and an equally interesting history. They are actually one of the oldest breeds of chicken. Records of the Silkie date back to at least the thirteenth century in Asia when they were first recorded by Marco Polo, though some ancient poetry from 700 CE appears to describe them as well. Because they were developed so long ago, their exact origin is unknown, but it is thought to be China, India, or Java.

The Silkie is instantly recognizable for its furry plumage, but it is also identified by its black skin, extra toe on each foot, walnut comb, and pom-pom. Some varieties also sport a beard or muff under the beak. They are among the broodiest breeds around, and some farms keep at least a few because they make such excellent mothers.

Silkies are an ornamental breed of chicken that has a tendency to seek out human companionship, and they are especially gentle with children. They are even known to "purr" when given affection by their favorite humans. Because of their compact size and inability to fly, Silkies are well suited to small backyards. They are, however, very vulnerable to predators and need extra protection. Silkies are also more susceptible to cold and wet weather due to their feathers' poor insulation qualities.

MARANS

The Marans (spelled with an "s") is a breed of chicken that originates from France. This is a wonderfully diverse breed that comes in a wide range of colors and patterns, including some with feathered legs and feet. If you are looking for a beginner-friendly, mild-mannered chicken that will take your egg basket to the next level, you can't go wrong with a Marans.

The Marans have been farm-favorites in their native France since at least the nineteenth century, though the original line of chickens used to develop the breed likely dates from far earlier. Marans were prized for their quality meat and their abundance of rich, dark brown eggs. They did not arrive in the U.S. until the early twentieth century, where they remained fairly obscure until much more recently.

Today, the Marans is a highly desirable backyard breed, prized for its uniquely dark eggs. For the darkest-colored eggs, the Black Copper Marans (see left), Wheaten Marans, Mystic Marans, and Blue Copper Marans are the best options. Actual egg color will vary, based on several factors, including the breeder, time of year, and the hen's age. Nearly all Marans will lay a darker egg than most other breeds, though, and are a wonderful addition to a backyard flock, especially if you are looking for a colorful nest of eggs to gather each day.

Chapter Two

❦

CHICKEN LIFE

THE CHICKEN LIFE CYCLE

Chickens have been used in classrooms throughout the U.S. and Europe for generations to study life cycles, and for good reason. Their well-documented stages of development and ease of breeding make them ideal for students to observe and understand the progression from egg to adult chicken.

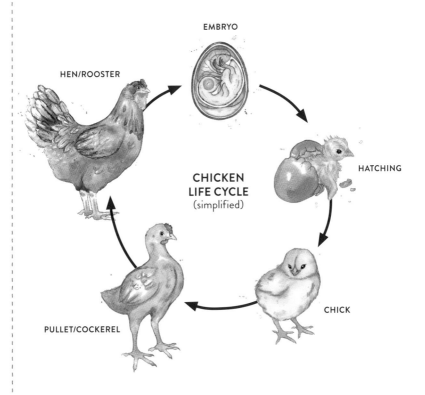

EMBRYO

HEN/ROOSTER

HATCHING

CHICKEN LIFE CYCLE
(simplified)

CHICK

PULLET/COCKEREL

EGG STAGE

Safely tucked up in its egg, a chicken embryo slowly develops under the warmth of its mother hen or in an incubator. The process from egg to hatching takes about 21 days. During this time, the developing chick is nourished by its yolk and fueled by small pockets of air that seep through the egg pores. If brooded under a hen, the chick's egg will be carefully turned and kept warm by its mother, who will dutifully sit on her clutch for over 23 hours a day, leaving only briefly to get food and water. Incredibly, the mother hen will also softly cluck to her eggs, teaching them to recognize her voice so that they will know her upon hatching. Within roughly 12–15 days or so of incubation, the embryos begin to resemble birds, with clear beaks, eyes, wings, and toes. On day 18, the chick positions itself to hatch.

HATCHING STAGE

Hatching is a very tiring, vulnerable time for a chick, with the whole process taking up to 24 hours. When ready to begin, the chick will use a special tip on its beak, called the "egg tooth," to slowly peck a small crack in the shell in an action called "pipping." During this time, you can actually hear the chick cheeping inside the shell. Once they begin to break through, it can take several more hours, or longer, for a chick to fully emerge.

CHICK STAGE

Once hatched, a baby chicken will be exhausted, but the yolk they consumed inside the egg can keep them nourished and hydrated for up to three days, if necessary. This gives the rest of their clutch time to hatch, which generally takes place within a span of two to three days. Unlike many other species, chicks are hatched nearly fully developed—they are able to see, walk, and eat on their own within just two to four hours of hatching. Chicks are, however, very vulnerable to cold and unable to regulate their own body temperature until fully feathered. Even though chicks are covered in a fluffy, downy coat upon hatching, they must be kept in a warm brooder, or protected and warmed by their mother. If raised by a mother hen, the chicks will follow her closely. For her part, the mother hen will dutifully protect her chicks from danger and teach them valuable skills, such as where to dust-bathe, and which food is tastiest to eat.

JUVENILE STAGE

Once their fluffy chick down is replaced by a full coat of feathers—usually at around seven to eight weeks of age—a chick enters the juvenile phase. They no longer require heat at this point, and if raised by a hen, they will start to become more independent from their mother, who may give them a little peck once in a while to encourage them to forage on their own. At this point, the differences between males and females start to become clear. Young roosters, called "cockerels," may begin to crow, and they will start showing larger red combs and distinctive pointed hackle and saddle feathers, as well as curved sickle feathers in the tail. The young hens, called "pullets" (see left), are still too young to lay or breed. At this stage, the females tend to have smaller, paler combs and more rounded feathers than their male counterparts. Juvenile chickens closely resemble their adult counterparts, but they are clearly smaller and slimmer, and a bit gangly-looking. Their "voices" also begin to change from the high-pitched "peep" of chicks to the throatier "cluck" of adults.

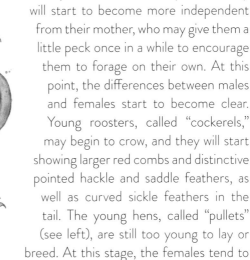

ADULT STAGE

Roosters and hens reach sexual maturity between 18 and 24 weeks of age, depending on the breed. Usually, production layers and smaller breeds will mature a little faster, while bigger breeds, like Orpingtons and hearty Brahmas, will mature a little slower. At this stage, cockerels—now fully developed with clear rooster plumage—will start attempting to court and breed with their hens, officially marking their move into their adult "rooster" stage. Right before laying, pullets will start to develop redder faces and combs, and will exhibit mating behavior called the "submissive squat," signifying their own move into adulthood. At this point, some keepers may begin calling them "hens," though others will continue to call them "pullets" until they reach one year of age.

COURTSHIP AND BREEDING STAGE

When a hen is ready to start raising the next generation of chicks, she will enter a phase called "broodiness," whether a rooster is present or not. Some breeds carry this brooding instinct much more strongly than others, and it isn't unheard of for some hens to never exhibit this behavior. Others may "go broody" multiple times a year. At this point, the broody hen will go through several major changes. She will lose the feathers on her chest and tummy, to better exchange heat to her developing eggs, and will sit on her nest for all hours of the day, emerging only once or twice to eat, drink, and relieve herself. She will also stop laying, and may become very aggressive, puffing up and growling angrily at anyone who approaches.

Broodiness is a very normal behavior, but it is extremely taxing on the hen, so it should be discouraged if there aren't any eggs to hatch or chicks to raise. If, however, she is sitting on a clutch of fertile eggs, the mother hen will dutifully protect and warm them until the next generation of chicks hatches, starting the life cycle anew.

EGGS

The incredible ability to lay a bountiful supply of eggs throughout the year is a special trait that is only exhibited by chickens and other poultry, and it is one of the greatest reasons for their popularity today as a food source.

EGG ANATOMY GLOSSARY

The humble chicken egg is truly a marvel all on its own, and it has some pretty fascinating features, each of which plays a vital role in supporting an embryo's development or protecting the egg until it is consumed or hatched.

YOLK

The yellow-orange yolk is the center of the egg. Made of amino acids, it is the most nutrient-dense part of the egg, and serves to feed the embryo as it develops into a chick. The color of the yolk may vary, depending on the hen's diet. In general, diets heavy in grain and soy produce lighter yellow yolks, while diets rich in insect protein or certain herbs like calendula will produce darker orange yolks.

EMBRYO

The embryo *only* exists in an egg if it is successfully fertilized, and it is rarely present in eggs sold for human consumption. If the embryo is present, it will be nestled safely alongside the yolk, and will eventually develop into a chick after 21 days of incubation.

CHALAZAE

The chalazae (singular, chalaza) are two spiral, rope-like strands of protein that help to suspend the yolk in the egg, preventing it from bumping up against the shell.

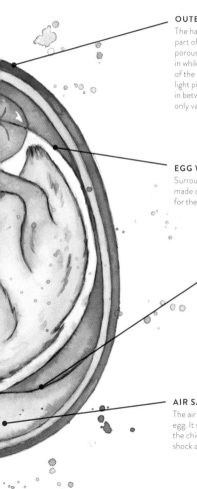

OUTER SHELL
The hard outer shell of a chicken egg is the last part of an egg to develop inside the hen. It is a porous structure that is meant to allow some air in while keeping bacteria out. The outermost layer of the shell may contain pigment, ranging from light pink to dark green, and many other colors in between. This pigment is purely genetic and only varies slightly throughout a hen's lifetime.

EGG WHITE
Surrounding the egg yolk, the egg white, made of albumen, provides a cushion for the developing embryo.

SHELL MEMBRANE
Between the internal egg and the hard outer shell are two thin membranes that help keep out harmful bacteria. Air is held between them, which expands as the egg ages. If an embryo is present, this pocket of air becomes vital for the chick to breathe as it begins hatching.

AIR SAC
The air sac is a pocket of air at the rounded tip of the egg. It serves several purposes, including providing the chick with air to breathe during hatching, shock absorption, and pressure regulation.

FUN CHICKEN EGG FACTS

1 A hen is born with all the possible eggs that she will lay in her lifetime.

2 The record for the most eggs laid in a single year by a chicken is 371, by a White Leghorn hen.

3 Most non-production breed hens will lay between 90 and 200 eggs in a year.

4 A hen's peak egg-laying time is between one and two years of age. After that, her egg production will slowly decrease, though many breeds are known to lay eggs well into six to eight years of age.

TRUE OR FALSE? DEBUNKING COMMON EGG MYTHS

Hens need roosters to lay eggs.

False! Contrary to popular belief, hens do not need roosters to be present in order to lay eggs. They will lay eggs whether they are fertilized or not.

Eggs sold in grocery stores are chick embryos.

False! Nearly all eggs purchased at grocery stores are sterile and have no chance of growing into a chick.

Fresh eggs are harder to peel.

True! Fresher eggs have less air inside them and tend to be harder to peel after boiling or steaming.

Eggs that sink in water are fresher than eggs that float.

True! Older eggs have more time for air to reach the inside of the shell, making them float. The higher they float, the older they are.

Brown eggs are healthier than white eggs.
False! The color of a chicken's egg shell is genetic and has no impact on the egg's nutritional content.

Small blood spots in an egg yolk means the egg is fertilized.
False! Tiny drops of blood in an egg yolk can occur if a blood vessel in the hen ruptures during the egg's formation. This is common and the egg is perfectly safe to eat.

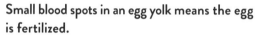

Eggs should always be refrigerated.
It depends! If the egg is washed, like they are in the U.S., they should be refrigerated. This is because washing removes the protective "bloom" (see left) coating on the shell, which if left intact keeps out harmful bacteria. Unwashed eggs are safe to store at room temperature for up to three weeks.

Intelligence and Behavior

Chickens are very busy and highly social animals. They spend most of the day wide awake, foraging, grooming, and vocalizing with one another. (The image above shows a chicken vocalizing with flockmates.) New keepers will notice that chickens are insatiably curious as well, and highly entertaining to watch. This behavior stems from the original wild Junglefowl species that chickens herald from, where survival depends on tightly knit social groups and a keen eye for finding both food and predators.

"BIRD BRAIN:" HOW SMART ARE CHICKENS, REALLY?

Chicken brains might be smaller than a pinto bean, but these creatures are surprisingly intelligent, especially regarding communication and socialization. In fact, they've been the subject of numerous studies on animal intelligence, and more is being discovered about their remarkable intellect every year.

THE PECKING ORDER

Chickens are deeply social animals that have a complex social structure known as the "pecking order." This social hierarchy is carefully maintained through communication among individuals, who may, if necessary, resort to fighting to establish their place in the group.

Generally, higher-ranking chickens will get the first pick on roosting places, dust baths, and choice food morsels. Lower-ranking chickens generally back down and keep out of the way. When conditions are healthy, even large flocks can maintain social order with very few scuffles or behavior issues, with chickens preferring to communicate with eye contact or a quick peck to settle disputes. This remarkable social structure is one reason why chickens were so easy to domesticate.

Outside settling the pecking order, chickens show remarkably empathetic and nuanced social behaviors too. They tend to form friendships with one or two individuals, with whom they will spend most of their time, foraging together and cuddling on roosting bars. Friends will even engage in "allopreening," gently cleaning one another's feathers on the head and face. In some cases, these tight-knit friendships will even extend to chickens helping one another if one is disabled. There is even some evidence that suggests that chickens grieve if a favorite flockmate passes away.

COURTSHIP

One of the most fascinating chicken behaviors to witness is courtship. Roosters, like many other male birds in different species, engage in elaborate courtship displays to their harem of hens. During spring and summer especially, a rooster will mate with his hens dozens of times a day. To ensure his hens accept him, a rooster will court them by puffing out his chest, crowing, "dancing" with his wing down, and stomping his feet. (See opposite for an example of a wing dance.) To prove that he's a good provider, a rooster will also call his hens to tasty morsels of food, rather than eating himself, in a behavior known as "tidbitting." In some cases, roosters will even help their hens find a suitable nesting place to lay their eggs.

Hens, for their part, are surprisingly picky about which rooster they allow to fertilize their eggs. Hens are known to show preference to roosters with bright red combs and loud crows, but they also seem to favor roosters who tidbit often. Even after mating, a hen has the ability to "expel" the sperm of any rooster who she deems unfit.

LITTLE RAPTORS:
THE DARK SIDE OF CHICKENS

Despite their complex social structure and ability to care for one another, chickens do not align with human morality, and they can be surprisingly violent if a particular situation presents itself. This can include being territorial with unknown chickens or other animals, when experiencing limited resources, or even in cases of illness or injury. Using sharp beaks and claws, they will attack the threat, chasing and even mortally wounding it if necessary. Roosters can be especially aggressive, and are known to cause injury to other species and humans, but hens will exhibit this behavior too.

In cases of stress or limited resources, chickens are also known to bully one another in their own flock, and will even drive out a flock member who they deem a liability, such as in cases of illness or injury. It is not out of cruelty, but out of necessity— the safety of the flock outweighs that of the individual, as far as chickens are concerned. This behavior can be so extreme, though, that it can on occasion have fatal consequences. Keeping the peace in a backyard flock means ensuring that new flock members are introduced slowly and carefully, and that the flock always has adequate space, food, and water. Isolating injured or ill chickens is also very important for this reason.

Contrary to popular belief, chickens are predatory, and they are not herbivores at all. Free-ranging chickens are opportunistic hunters, eating animals like spiders and grasshoppers, as well as mice, lizards, and even small birds. Chickens will also eat carrion and leftover meat, if these are offered. This behavior has earned them the nickname "fluffy raptors" among backyard chicken keepers. (The image below shows a squabble between members of a backyard flock.)

TRAINING AND MEMORY

Up until recent decades, not much was thought about chicken intelligence as a whole. Chickens are curious to a fault and nervous bordering on neurotic sometimes, which has earned them the unfair perception of being unintelligent. Upon closer inspection, however, chickens are showing backyard keepers and researchers alike just how intelligent they really are. (The image on the left shows an intelligent Barred Plymouth Rock with alert eyes and a curious expression.)

For starters, chickens have excellent memories and are remarkably trainable, quickly picking up on new behaviors in order to obtain a reward. They are so easy to train, in fact, that they are frequently used among professional animal trainers to perfect training techniques. Easy tricks that backyard chickens can learn include recall (coming when called), "spin," "jump," and "peck a target."

Recent studies also suggest that chickens are able to count up to about five, and have a sense of "less" versus "more" at the same level as primates and small human children. Even chicks demonstrated an ability to count by consistently picking a screen hiding a number of objects that they knew was larger. Interestingly, chickens have demonstrated the concept of delayed gratification, meaning they will forgo a quick reward and wait for an even bigger reward that they learned to expect. This behavior not only shows self-control, but also the ability to plan for the future.

CHICKEN INTELLIGENCE FACTS

1 Chickens have a keen sense of direction and are able to navigate over long distances back to their roosting place using landmarks.

2 Chickens can recognize up to 100 unique individuals and faces, including other chickens and different animals, such as dogs and cats. They can even distinguish between different people.

3 Chickens are highly social animals and should be kept in groups of two or more for optimal health.

4 Chickens use over 30 distinct vocalizations to convey messages to one another. They even have different "words" for ground predators versus aerial predators.

5 Chickens can teach one another new behaviors, such as where to find food—or how to sneak into the house!

Chapter Three

❧

RAISING CHICKENS

BENEFITS OF KEEPING BACKYARD CHICKENS

WHY KEEP CHICKENS?

Chickens come in behind only cats and dogs as the most common pets in the U.S. and U.K., with an estimated 85 million chickens kept in backyards. Their popularity is for good reason: chickens are fun, companionable animals, with the added benefit of producing delicious, nutritious eggs!

There is no true dichotomy between being kept as livestock or pets when it comes to chickens; their relationship with their keepers is a very personal decision and may vary widely. Some keepers may choose to give their birds the same level of care as other pets, including regular vet visits, grooming, and teaching them tricks. Others may prefer to view their poultry as beloved farm animals that provide high-quality, sustainable food. Still others may intend to keep chickens for their eggs but find to their surprise that they get attached to the birds. There is no "right" or "best" way to view backyard chicken-keeping, so long as the flock is provided with adequate care.

PETS WITH BENEFITS

There are many wonderful benefits to keeping a small flock of chickens in the backyard, as millions of households are quickly discovering.

Eggs

It's true that eggs fresh from happy, healthy backyard hens are better quality than store-bought. Hens that are given ample space to roam and forage, with access to dirt and sunshine, produce eggs that are higher in omega-3s, vitamin A, vitamin E, and beta carotene. Their eggs are also 30 percent lower in cholesterol and 20 percent lower in saturated fat. Backyard chicken eggs are also significantly fresher than store-bought, which average about 60 days old by the time they reach the shelves at grocery stores. The nutritional quality of eggs from backyard hens all comes down to diet and welfare: backyard hens that are not cared for properly will not produce more nutritious eggs.

Mental Health

Keeping chickens comes with a range of benefits outside their fresh eggs. One of the most often overlooked is their positive impact on humans' mental health. Studies have already established that pets can help reduce stress, anxiety, and depression, and chickens are no exception. The act of caring for them provides routine and structure, and has the added bonus of ensuring time is spent outside every day, which comes with health benefits of its own. Many keepers report that keeping chickens has helped them with PTSD, loneliness, and even addiction recovery. Some programs in recent years have even started to incorporate chickens as therapy animals for assisted-living facilities.

Sustainability

Chickens are a valuable asset to small homesteads and gardens, making even small city backyards into a more sustainable home, especially for houses with backyard gardens. Chicken manure is rich in nitrogen and makes excellent fertilizer, and the birds are highly efficient at turning kitchen scraps, garden weeds, and raked leaves into rich compost.

Although commonly viewed as gentle grain-eating birds, chickens are actually voracious predators with a keen taste for common garden pests, such as small rodents and bugs. A flock of chickens is therefore highly valued as natural, sustainable pest control, feeding on insects such as grasshoppers, grubs, spiders, and centipedes.

Education and Connection with Nature

It is hard to quantify, but the act of caring for livestock brings keepers of all ages closer to where food comes from and creates an interconnectedness with nature that cannot be replicated. Several studies on the interaction between children and livestock in particular have found that raising children with livestock, like chickens, is associated with physical, cognitive, and immunological benefits. Positive impacts documented include lower rates of asthma, and above-average emotional regulation.

Children and adults alike benefit from the educational experience of keeping chickens. Many schools and classrooms still practice hatching chicks each year to learn about life cycles, and some progressive schools have taken to keeping a small flock of chickens to teach about food production and animal welfare.

CHICKEN-KEEPING CHALLENGES

Although keeping a few chickens is relatively easy—many claim a small flock is easier to care for than a typical housecat—they are generally outdoor animals and therefore come with some unique challenges that new keepers should be aware of.

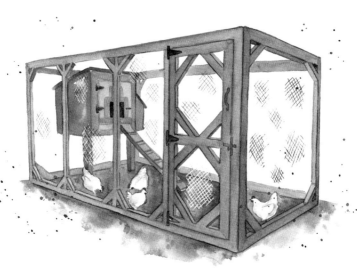

NOT HAVING ENOUGH TIME

Like any pet, it requires commitment to care for chickens properly. They can also live surprisingly long lives, up to 12 years or so. Backyard chickens require consistent care as well. Ideally, they should be checked on at least once daily to gather eggs, top up food and water, and to check for any illnesses or injuries. However, they are, fortunately, pretty self-sufficient animals that don't require a ton of time or attention and can be left alone in a secure enclosure with adequate food and water for a couple days or so if needed. A typical time-commitment guideline for a small chicken flock looks something like this:

1–2 x daily: 5–10 minutes. Open and close the coop, check the water and food, and gather eggs.

1 x weekly: 10–20 minutes. Clean the coop, rake the outdoor run, and scrub and refill food and water containers.

1 x monthly: 30–45 minutes. Deep-clean the coop and perform seasonal maintenance, such as winterizing the coop in winter or adding shade in the summer.

As-needed: Varies. Perform tasks such as providing basic medical care, expanding or repairing the coop, dealing with behavioral issues, or integrating new chickens.

For families with busy schedules, chicken chores are often a much-appreciated respite from the daily grind, and even young children can help with quick tasks like gathering eggs or refilling food containers.

NOT HAVING ENOUGH SPACE

Arguably the biggest challenge to keeping chickens is simply not having enough space for the flock that you want. As the saying goes, "chicken math is real!" A plan for a small, three- or four-chicken flock can quickly spiral into wanting a dozen or more birds. They are simply too fun to keep, and too adorable to resist as chicks each spring. A flock that is squeezed into a space

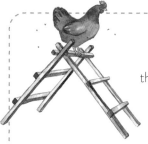

that is too small, however, will likely experience major health and behavioral issues, ranging from aggression to disease.

Generally speaking, a coop where chickens lay eggs and sleep should be large enough to ensure at least 2–6 square feet (⅕–½ square meters) of space per chicken, and a chicken run or outdoor pen should have 10–20 square feet (1–2 square meters) of space per chicken. In addition, chickens require 8–12 inches (20–24 centimeters) each of roosting bar space, and one nesting box per four hens.

Keep in mind, some chickens may require more or less room, depending on their size and energy levels. Some situations may also require additional space, such as a brooding pen for a hen with chicks, or a quarantine coop for new chickens. When in doubt, make sure to err on the side of more space to ensure a happier, healthier flock at any size.

PREDATORS

Chickens are perhaps the most highly sought-after prey animals among local predators. Even in city backyards, chickens are exceptionally vulnerable without adequate protection, and nearly all chicken keepers experience heartbreaking loss due to predation at one point or another. It can be especially devastating if the predator in question is the family dog or cat. In order to keep a flock safe, secure enclosures and safety practices are key. The specific type of predators will vary depending on your location—such as raccoons, bears, and coyotes in the U.S., or urban foxes in the U.K.

Predators can be classified as ground-dwelling or aerial, including the following:

Ground-dwelling	*Aerial*
Dog	Hawk
Cat	Owl
Raccoon	Eagle
Fisher or weasel	
Fox	
Coyote	
Wild cat or bobcat	
Bear	

In order to prevent unnecessary heartbreak, new keepers must assume that there are more predators in their area than they might know about. Domestic predators, like dogs and sometimes cats, can also be a bigger threat than most keepers realize. Even gentle, friendly dogs are likely to try to "play" with chickens, often leading to fatal results. Providing adequate protection—see Predator Protection Tips overleaf—in addition to a secure coop to lock chickens into at night, is key.

OUTDOOR CARE, NO MATTER THE WEATHER

One potential challenge to keeping chickens is that they are, in fact, outdoor animals that require daily outdoor time in order to thrive. When the weather is mild, it is a wonderfully enjoyable experience to head outside and feed chickens some treats in the warm sunshine. Unfortunately, chickens require just as much attention in poor weather, if not more so, which is not nearly as fun. That means having to gather eggs, refill water, and clean the coop, even in snow, rain, and high winds. For some, this can be a serious nuisance, especially in harsh climates. For others, however, having to go outside every day for fresh air no matter what is a great mental-health benefit (and still faster and easier to do than walking the dog).

For millions of households, the joys of chicken keeping far outweigh the drawbacks, but it helps to be aware of both the benefits and unique challenges of caring for these fascinating creatures. If the thought of getting chickens is still appealing, the next wonderful step is deciding which chickens to get!

PREDATOR PROTECTION TIPS

1 Build a fully fenced wire run around the coop so that the chickens can safely roam outside if predators are nearby. This also provides a second barrier from predators at night.

2 Use heavy-gauge hardware cloth, rather than "chicken wire" or netting, which are flimsy and easily broken by a determined predator.

3 Make the chicken run or pen dig-proof by burying wire fencing around the perimeter. Reinforce with rocks, pavers, or cinder blocks.

4 Add locks to the doors of the coop and the nesting box, especially if you have raccoons or foxes in the area, and always close the coop securely at night.

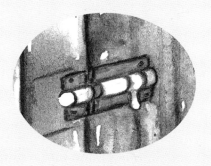

5 Use predator deterrents, such as motion-activated lights, streamers, or electric netting.

6 Offer low areas of protection, such as bushes or small shelters, that your chickens can hide under if a hawk or owl approaches.

7 If you have a family dog that is likely to chase, always follow the "two-barrier rule:" keep at least two closed doors or gates between the dog and the chickens.

8 Consider hiring a professional trainer to help make a dog safe around the chickens.

9 Do not leave chicks or chickens unattended with dogs, cats, or small children.

BANTAMS AND FRIZZLES AND ORPS, OH MY! A STEP-BY-STEP GUIDE TO CHOOSING CHICKENS

With such a dizzying array of breeds and colors to choose from, selecting the perfect backyard chicken can be a daunting experience. The good news is that, at the end of the day, a chicken is a chicken, and buyer's remorse is highly unlikely. Plus, no keeper needs to choose just one breed—a great way to enjoy many breeds is to select several at once for a small, varied flock.

FUN CHICKEN BREED FACTS

1 When Silkies were first introduced to Europe, they were sold as a cross between a chicken and a rabbit!

2 There are believed to be over 500 unique breeds of chicken in the world.

3 The color of egg that a hen lays is genetic—she will lay the same color throughout her life.

4 As with dogs, there are national chicken fancy clubs all around the world, complete with a formal "standard of perfection," breed registry, and judged shows.

5 Most non-industrial chicken breeds can live for between eight and ten years.

6 Many chicken breeds come in two sizes: standard, weighing 4+ pounds (more than 1.8 kilograms), and bantam, weighing under 2 pounds (1 kilogram). (The Cochin hens at the top of the page opposite show the Bantam size next to the standard size of the breed.)

7 Some chicken breeds are over 1,000 years old.

8 Not all chicken breeds are meant for meat or eggs. Some of the oldest breeds are prized for their looks or companionship.

GETTING CLARITY

In order to ensure your chickens are the perfect fit for your family, it helps to consider your primary purpose for getting chickens, and the resources you have to hand, including:

- Space available.

- Any noise, smell, or other possible concerns if the chickens are in close proximity to neighbors.

- Time or resource limitations.

- Children or other pets at your property.

- Climate.

If fresh eggs are your most important reason for getting chickens, consider purchasing a good egg-laying breed, like the Sex-links, or a heritage breed like the Rhode Island Red (see right). Colorful egg layers like the Easter Egger or Marans are also a great option for gathering or selling eggs. Avoid Silkies or most bantams, which do not lay as well or as often.

If companionship and cuteness is a major factor to getting chickens, and eggs are more of a bonus, then consider friendly

breeds like the D'Uccle, the Silkie (see right), and the Orpington. Avoid industrial layers, like Sex-links or Leghorns, which tend to have shorter lifespans, and less-friendly breeds, like the Polish or Sumatra.

If space and noise is a major concern, such as in a small city neighborhood, consider opting for docile, quieter breeds like the bantam Cochin, D'Uccle (see left), and Silkie, which require far less space than standard-sized hens, and tend to make less noise.

If you live in a more rural area and plan on free-ranging your flock or starting a small homestead, the dual-purpose heritage breeds are best, especially the Rhode Island Red and Barred Plymouth Rock (see right). Sex-links are also good options, and their low feed-to-egg ratio make them great assets to a homestead.

For homes with children, a docile nature often matters more than size when selecting chickens. Gentle giant breeds like the Orpington and Brahma are especially calm and sweet, and are less likely to be injured if a child squeezes them too tight

or drops them. Smaller breeds like the Silkie and bantam Cochin are also good options and generally tolerate being held and petted. Thanks to their colorful eggs, the Marans (see left), Easter Egger, and Ameraucana are often favorites among children as well.

For homes located in harsher, colder climates, cold-hardy breeds will be much less stressful and easier to look after. Consider getting large-bodied breeds with small combs, such as the Ameraucana (see right), Cochin, or Barred Plymouth Rock.

More delicate breeds such as some fancy exotic bantams and Silkies may require more protection than some keepers have the resources for if winters are especially harsh.

For homes located in very warm climates where heat is a concern, consider breeds like the Sex-links, Australorp (see left), and Rhode Island Red, which tend to not overheat as easily as fluffy breeds like Cochins or Orpingtons.

If you are unsure of your time commitment or are simply not sure exactly for what purpose you want chickens, a good rule to follow is to select one of the popular "heritage" breeds or other highly popular mixes. The heritage breeds, like the Orpington, Barred Plymouth Rock, and Australorp, have been favorites for generations, and are especially easy to find and keep, while the colorful egg-layer mixes, like the Easter Egger (see right) and Olive Egger, are fun and easy to care for no matter their primary purpose.

CHICKEN BREEDS TO AVOID

While most chicken breeds will happily suit most lifestyles and needs, there are some breeds that are far more specialized, and may not be the best fit for backyard flocks.

Broilers (a.k.a. Cornish Cross or Meat Birds)

These industrial meat-production breeds, commonly sold as Cornish Crosses in the U.S., are bred to grow as much meat as possible, as quickly as possible, before being slaughtered at 8–16 weeks. While they can be sweet-natured birds, they are medically fragile and do not often live much longer than six months before succumbing to organ failure or other issues. Unless kept to be a source of meat, these breeds are not recommended for backyard flocks.

Exotic or Fancy Breeds

There are dozens of breathtakingly beautiful chicken breeds available, ranging from the diminutive Serama to the long-tailed Onagadori, which have gained notoriety in the past several years. While these breeds can make great backyard pets, they may also require some specialized care that might not suit every family. Some of these exotic breeds require supplemental heat in winter, for example, or special housing to keep their feathers in good condition. Some exotic breeds are also prone to health or behavioral challenges, and may be better suited to experienced poultry-keeper homes. Poultry shows and breed-focused forums online are a great way to get acquainted with these more exotic breeds, and learn from the experts on whether or not they are a good addition to your flock. The Polish Frizzle (below) is an example of a fancy breed because of their uniquely curled feathers. They require special care to maintain, but they are a lovely and gentle breed.

Straight run: a random assortment of male and female chicks.

Pullet: a young female chicken or chick.

Cockerel: a young male chicken or chick.

Biddy: another word for chicks or young chickens.

Sexed male or female chicks: an assortment of only-male or only-female chicks. Sex identification in chicks is not perfect, and sexed chicks may still include unwanted males or females.

Off heat: older chicks that are fully feathered and no longer require supplemental heat.

Grow-outs: another word for pullets or cockerels. An assortment of fully feathered young chickens not yet matured.

CHICKS OR HENS?

Once a desired breed is selected, keepers will have a few other options to consider, such as choosing newly hatched chicks, adolescent pullets, grown hens, or even fertilized eggs. There are pros and cons to each of them.

Chicks

Chicks are frankly adorable and so hard to resist! Raising chickens from newly hatched chicks is the most reliable way to ensure tame, affectionate hens, and many breeds are more widely available as chicks than as adults.

Chicks are also very affordable, coming in at a fraction of the cost of an adult.

However, chicks do come with some challenges. They require specialized housing and care for the first 4–6 weeks and need to be checked on and cared for multiple times a day, usually requiring them to live indoors with a heat source. Busy families with travel plans may want to opt for easier, older birds instead. Chicks also come with the risk of being unwanted roosters, or an unwanted breed.

Pullets

Families that want to avoid the time commitment and care of chicks but still want to ensure they get young birds may opt for purchasing pullets. These young female chickens/older female chicks tend to range in age between 8 and 16 weeks. Pullets are fully feathered and don't require any special heat or housing, and by this age are very clearly female.

Pullets tend to be much more expensive than young chicks, however, and can be harder to find outside of large hatcheries or specialized breeders. Some breeds may not be available for purchase as pullets at all, and shipping can be costly. While pullets grow to be quite friendly, they tend to not become quite as tame as hand-raised chicks.

Hens

For keepers who are looking to get a laying flock quickly, or replace hens in their existing flock, fully grown hens are an excellent choice. Depending on their breed, age, and circumstance, fully grown hens can range from quite expensive to affordable and even free. In fact, many families

acquire hens from sanctuaries or from other keepers who need to rehome or downsize their flocks as a low-cost way to get started with chickens quickly. As with pullets, grown hens are easily identified, so there's no risk of unwanted roosters or getting the wrong breed.

Fully grown hens do come with some challenges, depending on where they are acquired. Hens rehomed from other farms or backyards may have unknown health issues or parasites, and they are very hard to age. A hen sold as being 1–2 years old may in fact be much older and past her laying prime. Buyer research, quarantine, and some flexibility is key when purchasing grown hens or roosters.

Fertile Eggs

Obtaining fertilized eggs, sometimes known as "hatching eggs," is a great way to get new chickens, and these are especially useful for keepers without roosters who have a broody hen eager to hatch chicks of her own. Hatching eggs can also be placed in an incubator,

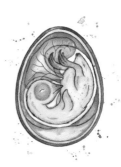

which is a fun process. Some breeders will ship hatching eggs, and this is sometimes the only way to get a rare breed. Starting out with fertilized eggs can be challenging, however—especially to start—and the process may include some sad moments if the incubation does not go well or a chick does not hatch properly. Fertilized eggs are also a random mix of male and female chicks, so rehoming is likely necessary if there are more roosters than wanted.

WHERE TO BUY CHICKENS

Thanks to their skyrocketing popularity, chickens are generally easy to find across any life stage, especially for the more common backyard breeds. As with dogs, cats, and other animals, the place you get your chickens from can vary significantly, with regards to the health, quality, and cost of the animals.

Hatcheries

Arguably the most popular way to get chicks each year is through one of many reputable hatcheries. Most of these hatcheries have online and mail-order catalogs available, which are usually released right after Christmas for the coming season, complete with breed information, quantities, prices, and even specific dates on which to have their chicks shipped. Generally, keepers will find the greatest variety of breeds this way, and the most popular are always easy to find.

Usually, hatcheries will only offer chicks or occasionally hatching eggs, which are shipped to the local post office for pickup. This can come with some limitations and challenges, including short delivery windows and the possibility of something going wrong during shipping. Be sure to only use reputable hatcheries that have passed NPIP or similar health certifications, and review their shipping and return policies carefully in advance.

Breeders

Some rare or new chicken breeds are not yet available at the large hatcheries. In these cases, a reputable chicken breeder is the way to go. Chickens purchased from breeders will be more expensive, but they'll also be of higher quality, with more transparency into their parentage and breeder flock history. Beware of some scammers and unscrupulous breeders—especially of "trendy" breeds—and review the breeders' NPIP certification status to ensure the birds are disease-free.

Feed Stores

There is perhaps nothing more exciting than coming home

from the local feed store with a box full of adorable fluffy chicks! Each spring through early summer, many local feed stores open up a section filled with chicks in a variety of popular backyard breeds. These chicks are typically shipped from a large hatchery, which the feed store then sets up in brooders. The cost per chick therefore tends to be a little higher than buying chickens directly from the hatchery; however, the chicks often have had time to adjust from shipping by being at the feed store, and it's very easy for keepers to select each individual chick that they want.

Rehomes or Poultry Swaps

Private rehomes and "poultry swaps" (markets where chicken keepers set out their chickens available for rehoming) are often the quickest and easiest way to get started with chicken-keeping, and are among the most reliable places to find grown hens. Private homes especially may be willing to part with their flock, along with all their equipment, for a low price, giving brand-new keepers a head start on their chicken-keeping journey.

Nothing is more frustrating or heartbreaking when starting out than when brand-new chickens come with major health or parasite issues, and unfortunately rehomed chickens and birds that come from poultry swaps or similar situations are more likely to experience health issues.

When purchasing chickens, be sure to do a thorough health inspection first:

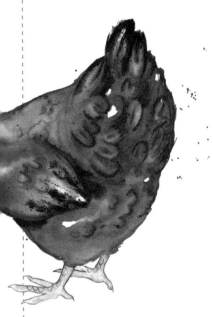

- *Check their legs.* Are the scales shiny and smooth? Or are there any signs of scaly leg mites (do the legs look rough, uneven, and swollen)?

- *Check their feathers*, especially by the vent (under the tail) and under the wings. Look closely for signs of small, black bugs, which is an indication of feather mites or other parasites.

- *Check their eyes and face*. A healthy chicken should have a clear face with bright eyes and smooth comb and wattles.

- *Check their breathing*. Watch out for any clicking or gasping, and for any signs of discharge from the nares (nose) and eyes.

- *Check their stature*. Are they alert and curious, or sleepy and listless? Be very wary of drooping tails, lethargy, and any sign of wobbling or limping, which could indicate serious illness.

A 14–30-day quarantine is highly recommended before introducing any new chickens to an existing flock, just in case there are hidden illnesses or parasites.

̓EQUIPMENT AND COOPS

Keeping chickens is a very front-loaded venture, in both effort and money. The equipment that a healthy, safe, backyard flock needs is not extensive, but getting them set up for the first time can be a serious investment. Setting chickens up the right way, however, is the best approach to ensure that chicken-keeping is as easy and enjoyable as possible, and it will save money in the long run.

WHAT CHICKENS REALLY NEED

The exact equipment that you need may vary from flock to flock, but a typical backyard flock should always have the following essential items:

1 A secure coop.

2 Nesting boxes.

3 Roosting bars inside the coop.

4 A secure fenced-in area (optional, but recommended).

5 At least one food container.

6 At least one water container.

7 A container for calcium supplements.

8 A container for grit (optional if coarse sand or fine rocks are available).

9 A dust bath area (optional if loose dirt outside is available).

10 Fresh food and water should be provided at all times.

Food and water containers and roosting bars are quite easy to find and purchase, but setting up a secure run and coop can be a little more daunting and costly. This is, however, essential to keeping a flock of chickens safe and comfortable year-round.

ESSENTIAL ELEMENTS OF A CHICKEN COOP

Arguably the most important piece of chicken-keeping equipment is the coop. It does not have to be fancy or pretty, but it should have everything the chickens need to stay comfortable and dry—especially at night—and safe from hungry predators. Exact designs and features vary widely, as does expense, but at their minimum, coops should always include the following:

1 An access door for the chickens that remains open during the day.

2 Nesting boxes for egg laying.

3 Roosting bars for sleeping at night.

4 A secure roof and solid walls that keep out rain, wind, and snow.

5 Ventilation holes or slots up high, near the roof line.

6 An access door for humans to gather eggs and keep the coop clean.

CHICKEN-KEEPING EXTRAS

In addition to the bare minimum, most keepers of backyard flocks use bedding, extra roosts, and other features to keep their birds more comfortable and to make chicken chores a little bit easier.

Bedding

The material used inside the coop and outside in the run varies by climate and needs. Some coops may not have bedding at all, and free-ranged chickens rarely need any kind of substrate outside. For most backyard keepers, however, the following materials are commonly used to help keep odors down and create a more comfortable space for their birds.

1. Sand (not play sand, which contains harmful dust particles).

2. Hemp.

3. Straw (not hay, which can mold).

4. Aspen or pine shavings (not cedar, which can be toxic to birds).

5. Loose dirt or soil.

Enrichment

Chickens are busy, curious creatures, and they greatly enjoy having new and interactive things to do during the day, especially if they are kept in an enclosed run. Providing extra enrichment activities for them can help prevent bullying among the flock and even reduce the amount of noise the chickens make. Some popular enrichment ideas include:

1 Outdoor perches and roosts.

2 Hanging food, such as skewers or vegetables, from the roof of the coop or run.

3 A mirror.

4 A swing—make sure it is big enough to hold a bird and is hung low to the ground for them to hop on and off easily.

5 Multiple feed and water containers—this can help to stop bullies guarding resources.

COOP EXTRAS

Outside the bare minimum, chicken keepers can also opt for several fun and functional features in their coop to make access or cleaning easier.

Windows. One popular feature is the addition of windows that can open and close. This helps keep the coop cooler in the summer and keeps odors down. For the chickens' safety, windows should be reinforced with hardware cloth.

Wallpaper and paint. While chickens do not likely care what their coop looks like inside, a nice coat of fresh paint or even wallpaper can help keep it looking clean, while having the added benefit of being easier to keep clean as well. Paint is also a great way to help prevent mites, which like to hide in small wood crevices.

Poop shelf. Chickens rarely spend time in their coop outside of roosting at night, so their droppings tend to get concentrated right below their roosting bars. A "poop shelf" is a convenient way to dispose of those droppings, making spot-cleaning the coop much easier. Poop shelves are set right below the roosting bars and are usually covered in a deep layer of sand for easy sifting with a kitty litter scoop. Poop shelves can also be a simple tarp that is emptied and sprayed down every morning.

Nesting box door. To help save time each day, a nesting box access door outside the coop is a quick and easy feature. Simply open the door each morning directly to the nesting boxes to gather eggs, and there's no need to enter the coop at all. This design does not work for all setups, though, and it's imperative that the nesting-box door can be securely locked and latched.

Insulation. For keepers in areas with cold winters, installing insulation can be a game changer to keeping their birds warm and comfortable. The insulation used can vary from reflective bubble wrap to spray foam, but it must be appropriately covered so the birds are not given access to it—they have a habit of pecking and eating it.

Run roof. A run covered in a solid roof is not only the best way to keep out predators, it also extends the coop in a way, by giving the flock a dry, comfortable area to forage in, even in the snow and rain. Covered runs also tend to be easier to keep clean. A run roof can be fancy and shingled, or just a simple tarp or roof panels. The weather-proofing effect can be enhanced by installing clear plastic tarps on the run walls as well, which is especially useful in winter to keep out wind and blowing snow.

Supplemental heat and light. Supplemental heat and light is a hot-button topic amongst chicken keepers, and for good reason. While there are benefits to including these, running electricity, especially heat, to a chicken coop comes with a heightened risk of fire, which unfortunately claims many coops and chickens' lives each winter. Heated dog bowls that are rated for outdoor use are generally a safe way to keep water from freezing in winter, and are a valuable asset in cold climates. Heating lamps in the coop, however, are significantly more dangerous and generally discouraged. Flat panel heaters, many of which have safety features to keep risk of fires low, are becoming a more popular alternative, but chicken keepers must be sure to do their research on the risks and benefits before using them.

Solar-powered lights are not overly risky for fire, but whether or not to use them comes down to preference. Supplemental light at night during the winter can help keep hens laying all year long, as their laying cycle is highly influenced by the hours of daylight. Many keepers, however, prefer their chickens not be given light at all in winter, to ensure their bodies take a healthy seasonal break from laying and extend the years that their chickens will keep laying.

MONEY-SAVING TIPS FOR BUYING CHICKEN EQUIPMENT

1 Instead of purchasing a large, expensive coop, get a cheaper used shed and upgrade it with roosting bars, ventilation, and nesting boxes.

2 Consider purchasing chickens and their coop all at once from families looking to move or who are stopping chicken-keeping.

3 Instead of building an expensive wooden run, consider purchasing a used metal dog pen and reinforce it with hardware cloth.

4 Avoid purchasing cheap online coops whenever possible, as they tend to break easily and are often too small for grown hens.

5 If you have the tools and time, download cheap or free coop plans online and use recycled materials.

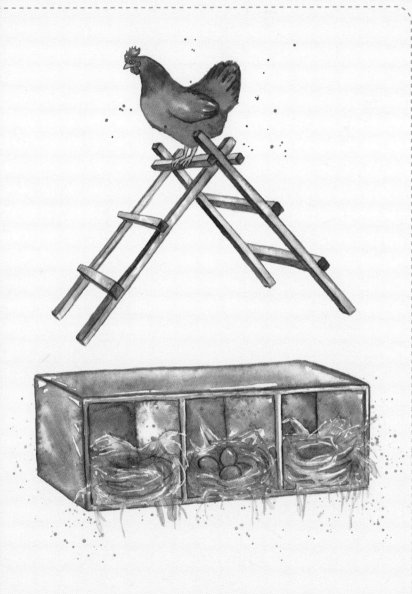

CONCLUSION

Chicken-keeping is a wonderfully rewarding experience, and one that is becoming more popular and accepted every year. As their popularity grows, chickens are challenging modern society's perception of them as purely farm animals, and the line between pet and livestock is getting increasingly blurred. No longer an afterthought at the grocery store, chickens and their eggs are proving to be a backyard staple for millions of households. For some, what started as a desire for fresh eggs has become a pursuit of a more sustainable life. For others, a small backyard flock has become a soothing place of companionship and entertainment to look forward to at the end of each day.

No matter the intended purpose of keeping chickens, keepers of all backgrounds quickly find that these fluffy, personable creatures help provide some respite from technology and the heavy schedule of life. But they come with some challenges, too, which many people are not used to experiencing—like carrying out chicken scratch during a blizzard, nursing a sick bird back to health, or experiencing the shock of a predator attack. The rewards are, more often than not, well worth the challenges, and chicken keepers everywhere are so happy you've joined us. Happy keeping!

ADDITIONAL
RESOURCES

ABOUT THE AUTHOR AND ILLUSTRATOR

ABOUT THE AUTHOR

Jessica Ford is a writer, mother, life-long keeper of chickens, and former competitor at American Poultry Association shows. She is also the co-author of *The Backyard Chickenkeeper's Bible* (HarperCollins*Publishers*), as well as the chicken and homestead contributor to *Home, Garden and Homestead* — an online "Guide to Modern Living" aimed at creating an independent, healthy, and sustainable homestead lifestyle. Jessica is also a self-professed "chicken nerd at heart." Jessica currently works as an author and digital marketing manager in Colorado Springs, U.S.A., where she lives with her husband, two children, and a small flock of chickens.

ABOUT THE ILLUSTRATOR

Amy Holliday is a freelance artist and illustrator based in Cumbria, England. Amy has worked on a wide variety of international projects, including children's and adult books, biological illustration, and packaging from food to beauty. For HarperCollins, Amy has illustrated *The Little Book of Bees* and *The Honey Book*. She works primarily in graphite and watercolors, before completing the piece digitally. She spends most of her days in her cozy home studio. Her favorite subjects to illustrate are all things related to the natural world. Other than art, Amy is passionate about wildlife conservation and environmentalism. She runs her own Etsy store, where she sells fine-art prints and products featuring her illustrations. You can find her online at www.amyholliday.co.uk.

BOOKS

The Backyard Chicken-keeper's Bible by Jessica Ford, Rachel Federman, and Sonya Patel Ellis, HarperCollins*Publishers*/ Abrams (2023)

Beautiful Chickens: Portraits of Champion Breeds by Christie Aschwanden, Ivy Press (2020)

The Chicken: A Natural History by Dr. Joseph Barber, ed., Ivy Press (2012)

The Chicken Encyclopedia: An Illustrated Reference by Gail Damerow, Storey Publishing (2012)

Chickenology: The Ultimate Encyclopedia by Barbara Sandri and Francesco Giubbilini, Princeton Architectural Press (2021)

WEBSITES

Backyard Chickens: www.backyardchickens.com

Backyard Poultry: www.backyardpoultry.iamcountryside.com

The Cape Coop Farm: www.thecapecoop.com

Community Chickens: www.communitychickens.com

The Happy Chicken Coop: www.thehappychickencoop.com

Hobby Farms: Chickens: hobbyfarms.com/animals/poultry

Homesteading: Raising Backyard Chickens:
www.homesteading.com/raising-backyard-chickens

iChicken: www.ichicken.ca

My Pet Chicken: www.mypetchicken.com

APPS

Cluck-ulator:
play.google.com/store/apps/details?id=com.chickenwaterer.
p5618jj&hl=en_US&gl=US

FlockPlenty: Chicken Egg Tracker
apps.apple.com/us/app/flockplenty-chicken-egg-tracker/
id1017524534

ORGANIZATIONS

Chicken Run Rescue: www.chickenrunrescue.org

Compassion in World Farming, Inc.:
"The Better Chicken Initiative": www.ciwf.com

HenPower: www.equalarts.org.uk/our-work/henpower

Penelope's Place—The Sanctuary:
www.penelopesplacethesanctuary.com

Safe Haven Farm Sanctuary:
www.safehavenfarmsanctuary.org/our-animals/chickens/

Triangle Chicken Advocates: www.trianglechickenadvocates.org

INDEX

ACKNOWLEDGMENTS

Jessica would like to thank Caitlin Doyle and the team at HarperCollins U.K. for this delightful project! She would like to extend a warm thank you to Amy Holliday for her incredible talent and charming illustrations throughout this book.

Amy is so grateful to the loveliest editor Caitlin Doyle, for approaching her with the idea of a second "Little Book..." and then surprising her three years later(!) with the exciting news of making it a reality. Also for her encouragement, patience, and her consistent enthusiasm for her work. Thank you to Amy's wonderful parents, for your endless support, encouragement and for never tiring of seeing her silly doodles. And to Amy's partner, thank you for enduring all the chicken talk, and for reminding her to take time off.

The publisher would like to thank both Jessica and Amy for being such a pleasure to work with as true and inspiring collaborators throughout. Thank you to Jacqui Caulton and e-Digital for the charmingly lovely design, and to the team behind the chickens: Rachel Malig, Jayoti Shah, and Chris Wright.